OUT OF THIS WORLD

Meet NASA Inventor John Mather and His Team's

Exoplanet-Spying Supertelescope

www.worldbook.com

World Book, Inc.
180 North LaSalle Street
Suite 900
Chicago, Illinois 60601
USA

For information about other World Book publications, visit our website at www.worldbook.com or call 1-800-WORLDBK (967-5325).

For information about sales to schools and libraries, call 1-800-975-3250 (United States), or 1-800-837-5365 (Canada).

© 2024 (print and e-book) by World Book, Inc. All rights reserved. No part of this publication may be reproduced, stored in a retrieval system, or transmitted in any form or by any means (electronic, mechanical, photocopying, recording, or otherwise) without written permission from World Book, Inc.

WORLD BOOK and the GLOBE DEVICE are registered trademarks or trademarks of World Book, Inc.

Produced in collaboration with the National Aeronautics and Space Administration (NASA).

Library of Congress Cataloging-in-Publication Data for this volume has been applied for.

Out of This World
ISBN: 978-0-7166-6564-9 (set, hc.)

Exoplanet-Spying Supertelescope
ISBN: 978-0-7166-6566-3 (hc.)

Also available as:
ISBN: 978-0-7166-6574-8 (e-book)
ISBN: 978-0-7166-6582-3 (soft cover)

Staff

Editorial

Vice President
Tom Evans

Senior Manager, New Content
Jeff De La Rosa

Writer
William D. Adams

Editor
Emma Flickinger

Curriculum Designer
Caroline Davidson

Proofreader
Nathalie Strassheim

Indexer
Nathaniel Lindstrom

Graphics and Design

Senior Visual Communications Designer
Melanie Bender

Digital Asset Specialist
Rosalia Bledsoe

Acknowledgments

Cover	© Lukasz Pawel Szczepanski, Shutterstock	23	© Best-Backgrounds/Shutterstock
		24-25	© Vadim Sadovski/Shutterstock
3	© Jurik Peter, Shutterstock; © Dorde Masovic, GrabCAD	26-27	Mason Media Inc./GMTO Corp./ASU
4-5	© givaga/Shutterstock	28-29	© Zakharchuk/Shutterstock
6-7	© Jurik Peter, Shutterstock	30-31	ESO/L. Calçada
8-9	John Mather; Herval (licensed under CC BY 2.0 DEED)	32-33	© KBDS/Shutterstock; NASA/JPL
10-11	© Dotted Yeti/Shutterstock	34-35	NASA/Chris Lynch
13	John Mather	36-37	NASA; WORLD BOOK
14-15	NASA/COBE Science Team; NASA	38-39	© Abner Gómez, GrabCAD
16-17	© Brian Maxwell, Shutterstock	40-41	© Dorde Masovic, GrabCAD
19	© Nobel	42-43	NASA/JPL-Caltech/T. Pyle
20-21	NASA/JPL-Caltech	44	John Mather

Contents

- **4** Introduction
- **8** INVENTOR FEATURE: Son of a scientist
- **10** Seeing a firefly next to a lighthouse
- **12** INVENTOR FEATURE: Mapping the CMB radiation
- **16** Occultation
- **18** INVENTOR FEATURE: The Nobel Prize
- **20** BIG IDEA: Starshade
- **22** INVENTOR FEATURE: The James Webb Space Telescope
- **24** The delicate—but resilient—Webb
- **26** The next generation of ground-based telescopes
- **28** BIG IDEA: Hybrid design
- **30** Adaptive optics
- **32** Staying in alignment
- **34** Getting to the next orbit
- **36** BIG IDEA: Contests
- **38** Contest winner
- **40** Other ideas
- **42** Future plans
- **44** Filling in the crossword puzzle of the universe
- **45** Glossary
- **46** Review and reflect
- **48** Index

Glossary There is a glossary of terms on page 45. Terms defined in the glossary are in boldface type that **looks like this** on their first appearance on any spread (two facing pages).

Pronunciations (how to say words) are given in parentheses the first time some difficult words appear in the book. They look like this: pronunciation (pruh NUHN see AY shuhn).

Introduction

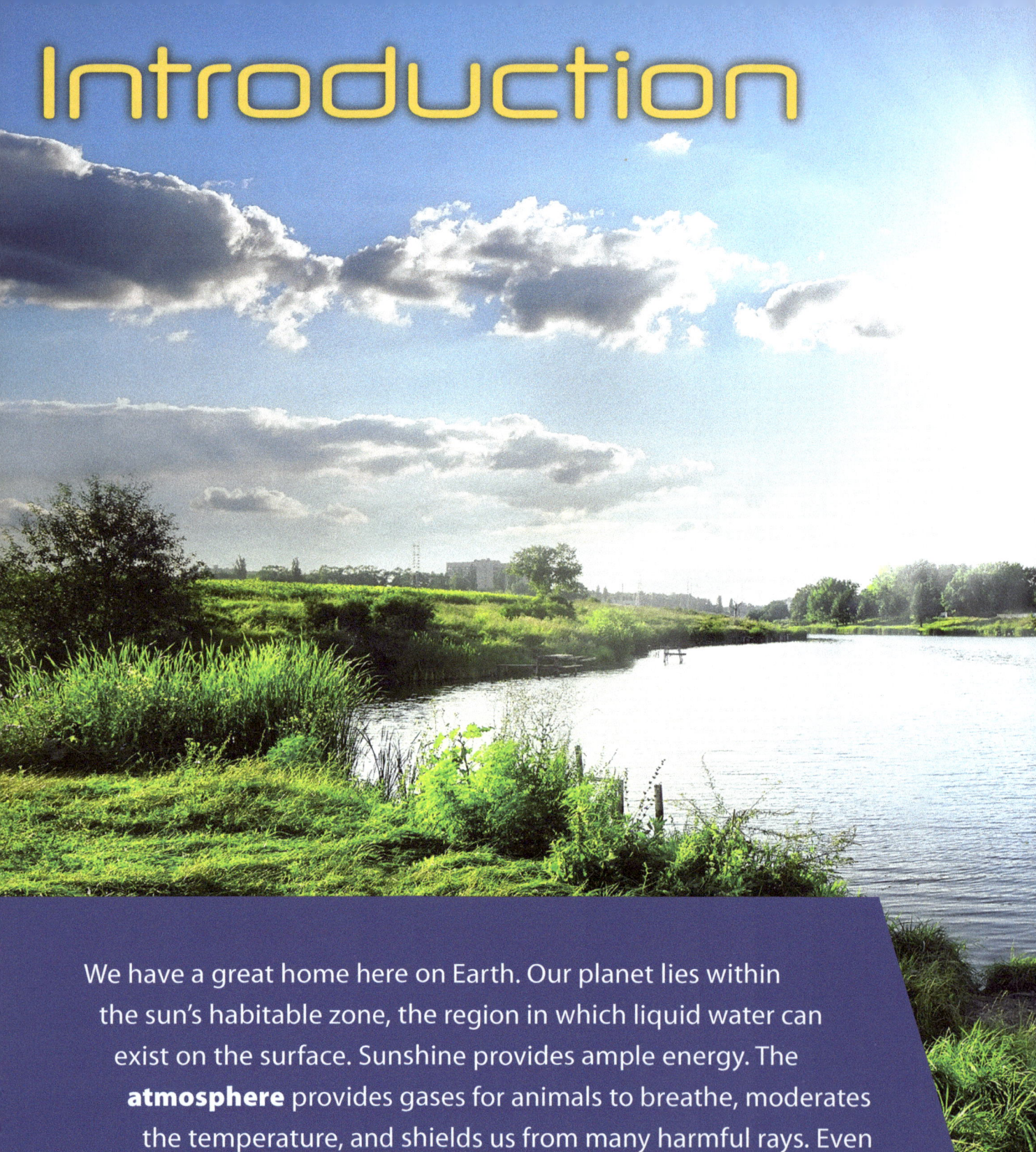

We have a great home here on Earth. Our planet lies within the sun's habitable zone, the region in which liquid water can exist on the surface. Sunshine provides ample energy. The **atmosphere** provides gases for animals to breathe, moderates the temperature, and shields us from many harmful rays. Even the **gravitational pull** of the giant planet Jupiter may help to prevent disastrous **asteroid** and **comet** impacts.

❝ But, is Earth so special and different that there are no other life-bearing planets elsewhere? ❞ —John

The nearest **exoplanets** (planets orbiting distant stars) are light-years away. A light-year is the distance light can travel in one Earth year, about 5.88 trillion miles (9.46 trillion kilometers). John Mather is developing a way to observe exoplanets despite the vast distances that separate them from Earth.

The satellite John is developing will be an orbiting shade, designed to block out the light from a star. With the glare from the star out of the way, a telescope will be able to capture images of the exoplanets orbiting around it. With such images, we can begin to understand their makeup and get the first hints as to whether any of them might contain life.

Artist's rendering of the surface of an exoplanet

The NASA Innovative Advanced Concepts program. The titles in the *Out of This World* series feature projects that have won grant money from a group formed by the United States National Aeronautics and Space Administration, or NASA. The NASA Innovative Advanced Concepts program (NIAC) provides funding to teams working to develop bold new advances in space technology. You can visit NIAC's website at www.nasa.gov/niac.

Meet John Mather.

❝ For decades, I've designed and managed radical experiments in space to improve our understanding of the universe. Now, I'm developing a special satellite that can improve the capabilities of giant ground-based telescopes. It will help us answer one of the main questions in astronomy: Are we alone in the universe? ❞

Inventor feature:
Son of a scientist

John grew up on a research farm in northern New Jersey. John's father was a scientist, working to measure and increase the protein content of milk. He told young John about how people are made up of cells and how each one contains genetic structures called chromosomes. John went on to study astronomy and physics, but cell biology has remained a passion of his.

> ❝ I'm fascinated to read about the discoveries of subcellular structures and how genes and chromosomes work. It's been on my mind all this time. ❞ —John

When John was young, his father took him to the American Museum

of Natural History in New York City. John was fascinated by the dinosaurs and other prehistoric animals, the planetarium shows, and the giant Cape York and Willamette meteorites. From that point on, John knew he wanted to be a scientist.

A young John with his sister (left), and the Willamette meteorite (right)

❝ I read books and played around with nuts and bolts and vacuum tubes and a few other things as a kid trying to learn science. ❞ —John

Seeing a firefly next to a lighthouse

An **exoplanet** is a planet that orbits a star other than the sun. The term *exoplanet* is short for *extrasolar planet*. Our galaxy, the Milky Way, probably has at least 100 billion and as many as 1 trillion exoplanets. Earth is undoubtedly special, but there are so many exoplanets that many astronomers suspect that life must also exist elsewhere.

Can we simply observe exoplanets with giant ground-based observatories and orbiting telescopes? Unfortunately, it is not that easy.

How do we know there are exoplanets if we cannot see them? Although astronomers usually cannot directly image exoplanets, they can observe their effects on their parent stars. A star might wobble ever so slightly, revealing the tug of an orbiting planet's **gravity.** Otherwise, a star's brightness might dip briefly as a planet passes in front of it, blocking out a tiny portion of its light.

Stars are extremely bright objects. Planets, on the other hand, give off no light themselves. Rather, they simply reflect light coming from their parent star. The difference in size and brightness is so great—a factor of around 10 billion—that it is like trying to see a firefly next to a lighthouse from miles or kilometers away. Just as the light from the lighthouse would make it impossible to see the firefly, the light from a star blinds telescopes to any nearby exoplanets.

Inventor feature:
Mapping the CMB radiation

In 1968, John went to the University of California, Berkeley, to obtain his Ph.D. degree. The American physicists Arno Penzias and Robert W. Wilson had recently made an amazing discovery about the universe.

❞ The cosmic microwave background [CMB] **radiation** had just been discovered a few years before. The CMB radiation is the leftover heat of the very early universe, from when it was very young and very hot. ❞ —John

For his Ph.D. project, John tried to map the CMB radiation from a ground-based or balloon-based observatory. But his experiment failed.

❞ I was lucky and they let me out of school with a thesis about a project that didn't work! ❞ —John

John as a young scientist poses with his parents.

But not all failures are bad, particularly in science. As luck would have it, John got a chance to try his experiment again.

> I went off to NASA thinking I'd do something else. But, I'd been there for about six months and NASA announced they wanted new proposals for satellite missions—this was just five years after the moon landing. So I said, 'Boss, my thesis project failed; we should try it in outer space.' He said, 'Yes!' —John

It was not just a lucky break. John knew the importance of this research, even though it did not work the first time. He kept at it to make sure it happened.

Inventor feature:
Mapping the CMB radiation cont.

NASA accepted the idea of developing a satellite to map the CMB **radiation.** John worked with **engineers** and other scientists to design the satellite, called the Cosmic Background Explorer (COBE). It launched in 1989.

First, COBE confirmed the temperature of the radiation. The temperature closely matched the heat output of a hypothetical object called a *blackbody*. Such an object absorbs all the radiation that reaches it and must therefore emit radiation itself to remain in equilibrium with its environment. This finding strongly supported the

Artist's conception of the Cosmic Background Explorer (COBE) satellite

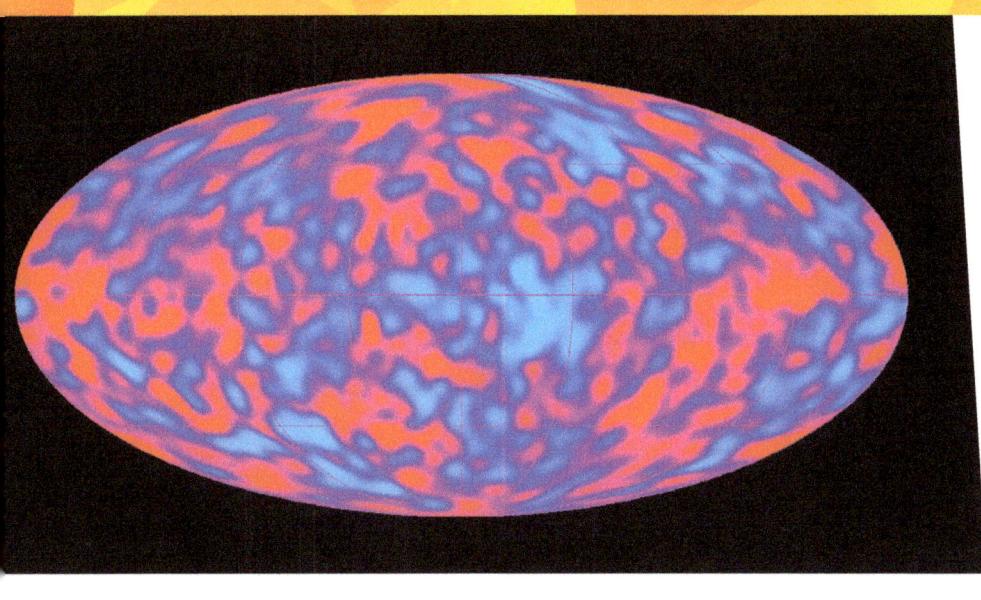

COBE recorded the data of the CMB radiation to make this map. The different colors represent minute temperature differences.

big bang theory that the infinite universe is expanding into itself, with no center and no edge.

❝ The conclusion from that was, yes, the expanding-universe story is correct. ❞ —John

COBE also discovered that there are tiny variations in the CMB radiation across the sky, like ripples spreading across the universe.

❝ Those spots explain why some parts of the universe were able to stop the expansion of the material and pull that material back in to form galaxies and stars and planets and people. We are here because of that. ❞ —John

Occultation

The sun and the moon have approximately the same apparent size in the sky. So, when the moon is in the path of the sun, it can block the sun's entire face. The sky darkens, sometimes revealing stars in the daytime. Such an event is called a *solar eclipse*.

Solar eclipses have been important in the study of the sun and physics. The 1919 total solar eclipse revealed how the sun's **gravity** bent the light of stars visible near it, supporting Albert Einstein's then-new theory of general relativity. Solar eclipses allowed *heliographers* (scientists

Solar eclipse as seen in North America, 2017

who study the sun) to observe the *corona*. The corona is the outer **atmosphere** of the sun. It is usually invisible due to the sun's intense light.

The solar eclipse is just one kind of *occultation*. An occultation is the hiding of the light of one star or other body behind another. The moon is often the occluding body, as it is during a solar eclipse. But the occluding body could be another planet or planetary moon, or even an artificial object. Astronomers have long sought to take advantage of the occultation effect to study other stars.

17

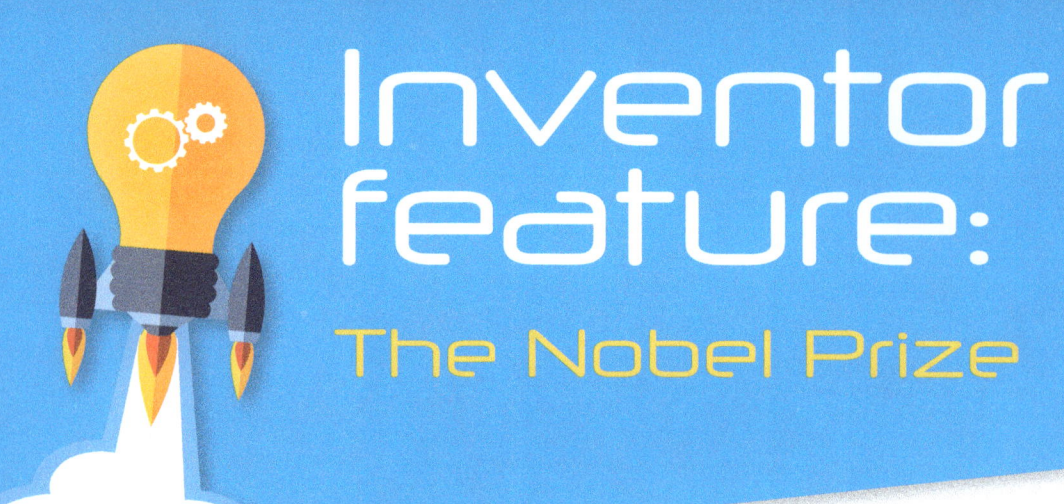

Inventor feature:
The Nobel Prize

The famous British physicist Stephen Hawking called COBE's discovery of ripples in the universe "the most important discovery of the century, if not of all time." Other scientists agreed with Hawking's assessment. About 6,000 scientific papers cited COBE's map—almost as many citations as there were pixels in the map itself.

Eventually, John received the ultimate scientific award for his work. In 2006, the Swedish Academy of Sciences awarded the Nobel Prize in Physics to John and the American astronomer George Smoot for their work with COBE.

❝ It was a thrill to go off to Stockholm and to see that my graduate student thesis had finally been done correctly and that everyone could see it! ❞

—John

Big idea:
Starshade

> ❝ Of course, I didn't invent **starshades.** They were invented by other people. ❞ —John

In 1962, the astronomer Lyman Spitzer proposed using an occulting disk—a circle of artificial material—to block the light of a star so its planets could be seen. But, major challenges caused the idea to be shelved for decades. When light hits the edge of a material, it *diffracts,* or scatters, around the barrier. The diffracted light could ruin the view of a telescope using an occulting disk. Therefore, the disk would have to be large to reduce the glare to the telescope caused by diffraction. But, a large occulting disk would have to be far away from the telescope so it would not block out the very planets the telescope was trying to observe. In addition, a circular disk scatters some of the diffracted light toward the telescope, anyway.

Webster Cash, an astrophysicist at the University of Colorado, Boulder, worked to create a shade that scattered diffracted light away from the telescope as part of a NIAC grant. After trying many designs, he discovered that a disk with petallike projections would scatter the light away from the telescope to reduce glare. Cash called the design a **starshade.** It would still have to be large and stationed far away from its telescope, but combined with other technological advances, the idea now seemed possible.

Inventor feature:
The James Webb Space Telescope

> **After we were done working on the Cosmic Background Explorer satellite, I thought, what am I going to do now? That was so exciting! We'll never do anything that good again.** —John

How does one follow up on Nobel Prize-worthy work? Not long after John wrapped up his research with COBE, he got a call inviting him to work on NASA's next-generation space telescope, the James Webb Space Telescope (JWST). John immediately accepted, and the next chapter of his scientific career began.

> **I've worked with scientists to decide what we really needed to build and engineers to see how we can build it.** —John

John is now the senior project scientist emeritus for JWST. He helps to pick, schedule, and prioritize the JWST's observation targets.

The delicate— but resilient— Webb

The James Webb Space Telescope is an astounding feat of precision engineering. It was originally planned to launch in 2007. But, the schedule and cost ballooned due to the project's ambitious scope. The next generation of space observatories required a mirror that was too large to fit in a conventional rocket **payload bay.** The telescope also needed a huge, multilayered heat shield to keep its instruments near absolute zero, the coldest possible temperature. JWST also had to be stationed thousands of miles or kilometers away from Earth, with no chance of being repaired by astronauts if something went wrong.

The Webb finally launched at the end of 2021. As it reached

> **"** I love the colleagues that I work with because they took this wish that astronomers had and they made real hardware that does the right thing. **"** —John

its destination, all of its delicate mechanisms *deployed* (spread out) properly. JWST has already delivered breathtaking images of the **solar system** and universe. Hundreds of papers have already been published using the data, with thousands or tens of thousands more likely to come.

As the value of the **starshade** became apparent, Cash pitched his idea to John. The two, along with others, lobbied NASA for it to be included with the JWST. They developed a scale-model **prototype** at the Jet **Propulsion** Laboratory. But the starshade was too technically challenging to add to an already-difficult JWST mission.

The next generation of ground-based telescopes

Orbiting telescopes are not the only way we can learn about our universe. It is easier to build telescopes here on Earth, and they are just as important in astronomy. The next generation of ground-based telescopes are under construction in the South American country of Chile.

The Giant Magellan Telescope is coming together at the border of the Coquimbo and Atacama regions. Seven circular mirrors will comprise the primary reflector, the surface that gathers light from the stars. These mirrors will be among the largest ever produced, each with a diameter of 27 ½ feet (8.4 meters). First light—the telescope's earliest observation—is expected in the late 2020's.

About 300 miles (500 kilometers) to the north, an even larger telescope is taking shape. For the appropriately named Extremely Large Telescope (ELT), almost 800 hexagon-shaped mirrors will make up a reflector 128 feet (39 meters) in diameter. First light for the ELT is planned in 2027.

An artist's impression of the Giant Magellan Telescope at the Las Campanas Observatory in Chile's Atacama Desert

Why Chile?

Chile has a population of about 20 million, with close to 90 percent of people living in or near cities. Therefore, large sections of the country have very few residents. These areas have very little light pollution and air pollution, both of which can interfere with astronomy. The country is lined with mountains and dry deserts. Positioning telescopes on mountains gets them above the thickest layers of the **atmosphere.** With dry air and few cloudy nights, telescopes can maximize their observation time.

Big idea:
Hybrid design

Designing, building, and launching the James Webb Space Telescope (JWST) was a monumental and incredibly complex undertaking. Still more powerful space telescopes are in the works, but they will be at least as complex as the JWST. Sending a **starshade** with one of them—a complex system in itself—seems daunting.

Space telescopes are always limited by the size of the rockets that can transport them. **Engineers** can devise tricks to cram in a larger reflector, as they did with JWST and will do with future space telescopes. But, they will never be able to approach the size of the colossal ground-based telescopes.

The insurmountable size advantage that ground-based telescopes have on space telescopes got John thinking.

❝ What if you could use a starshade in space with a telescope on the ground? ❞ —John

The key is that telescopes with larger reflectors can make observations

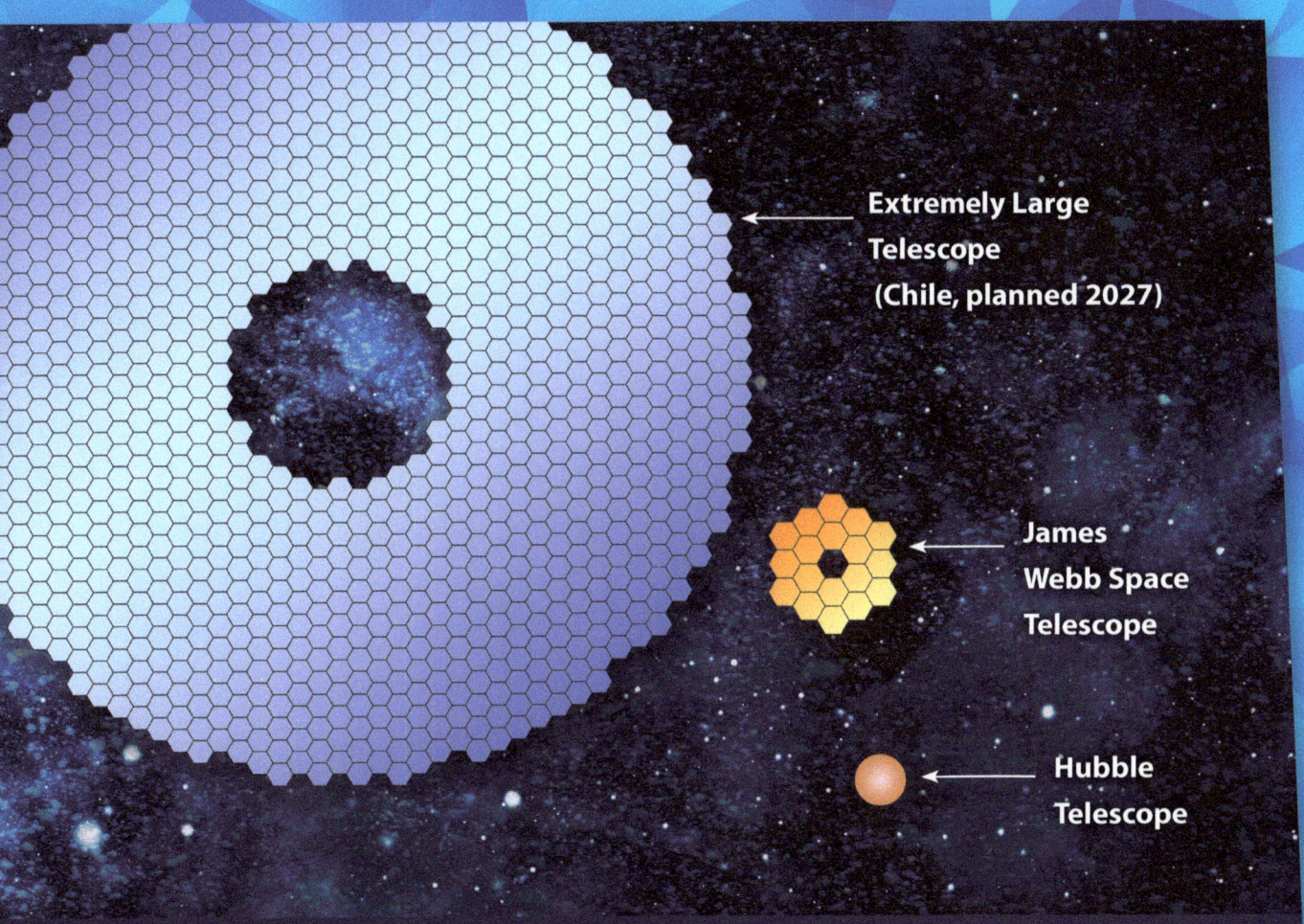

of distant objects faster than those with smaller reflectors.

❝ In fact, a telescope on the ground 39 meters across, like the ELT, could observe in one minute what it would take more than a day to observe with a 6-meter telescope in space. ❞ —John

John calls this plan the Hybrid Observatory for Earth-like **Exoplanets** (HOEE). The project is *hybrid* because it combines the space-borne starshade with a ground-based telescope.

Adaptive optics

❚❚ The big challenge for using a telescope—to observe visible light, anyway—is the **atmosphere** is turbulent. ❚❚ —John

The air is choppy, disturbed by wind and other fluctuations as the light passes through it.

❚❚ So it messes up the picture. You get a blurry picture when you wanted a sharp one. The idea is to focus the telescope on something that looks like a star by adjusting mirrors in the telescope to compensate for the turbulence. ❚❚ —John

Such an approach is called *adaptive optics*. Basically, a **laser** attached to the telescope acts as a fake star. By adjusting to atmospheric distortions in the laser light, the telescope can compensate for the atmosphere's effects on light from the stars. Earth-based telescopes already use adaptive optics—that's why you might have seen pictures of observatories firing lasers into the sky.

As a continuation of this idea, the HOEE **starshade** will carry a small laser and shine it down on the ground-based observatory when the two are in alignment. This will function as a low-light, artificial star that the telescope can use to precisely tune its mirrors to account for atmospheric disturbance.

" It's pretty hard to do in practice, but it's not impossible. " —John

This artist's rendering shows the Extremely Large Telescope in operation on Cerro Armazones in northern Chile. The telescope is shown using lasers to create artificial stars high in the atmosphere.

Staying in alignment

The HOEE **starshade** will be deployed in an *eccentric* (oblong) orbit around Earth. The orbit will have an *apogee* (the point farthest from Earth) of 105,000 miles (170,000 kilometers). Like a rollercoaster at the top of a hill, it will orbit slowly when it is far from Earth. Then, it will speed up as it plunges back toward the planet.

Near its apogee, the HOEE starshade will orbit Earth at approximately the same speed at which Earth rotates. So, it will appear to stay at a roughly steady point above the ground. At

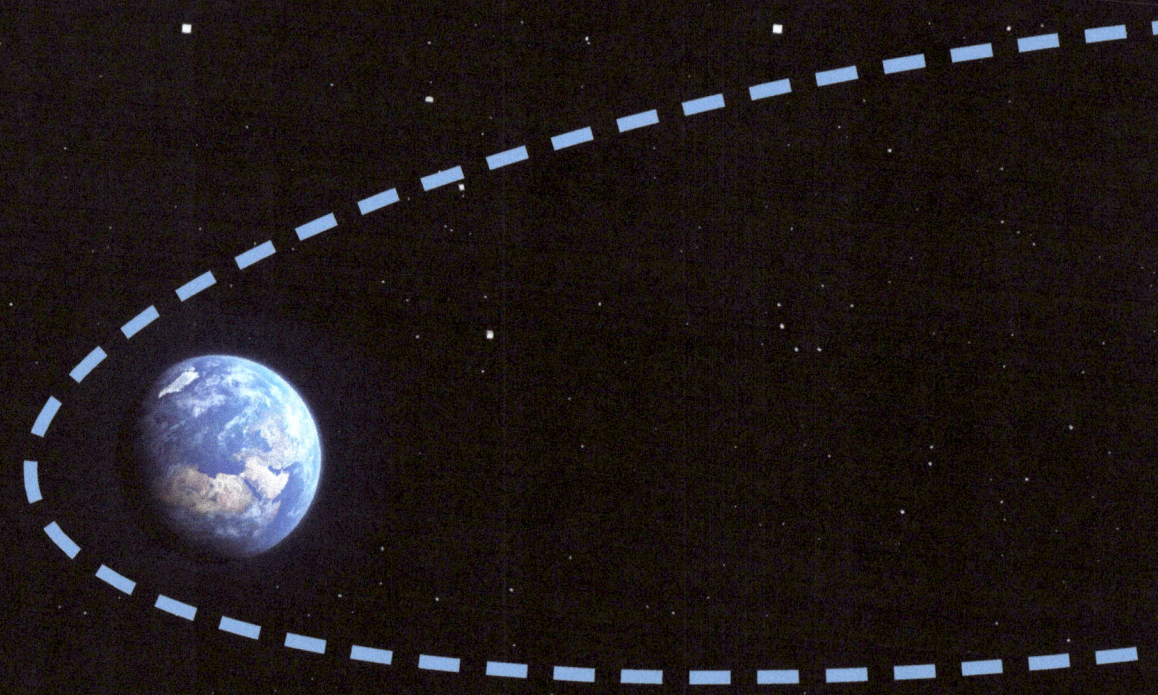

that point, the starshade will block the light of a target star from reaching a giant ground-based telescope. Such an alignment will last only for a few minutes, however, so the telescope will have a short time in which to capture pictures. The starshade will use rocket engines to extend this viewing time as much as possible. **Thrusters** will speed up or slow down the starshade to hold it in place as long as possible.

Getting to the next orbit

> **You get a small amount of force, but you can push for a really long time.** —John

Test fire of an ion engine

It would not make much sense to study the same star system over and over again. Therefore, the **starshade** will have to propel itself to different orbits. Conventional rockets deliver lots of **thrust** over a short time. But they use lots of heavy **propellant** to do so. The HOEE starshade will have many weeks to move into a new orbit as it follows its long path around Earth. But it will not be able to carry large tanks of heavy, conventional propellant. Therefore, John is designing his project with a second form of **propulsion** for repositioning the starshade—ion propulsion. In ion propulsion, a fuel is sent through an electrified grid. The grid heats the fuel and *ionizes* (electrically charges) it. The ions then exit through the nozzle at a high velocity, pushing the craft in the opposite direction.

The ion propulsion will be active for days at a time, slowly pushing the starshade into a new orbit for its next target.

Big idea:
Contests

An orbiting telescope would need a **starshade** about 160 feet (50 meters) in diameter. The HOEE starshade will need to be double that diameter for ground-based telescopes to get enough observation time during each pass. It will also have to be highly maneuverable to properly align with ground-based telescopes. A scaled-up version of existing starshade hardware would be too large to fit into existing rockets. Such difficult engineering projects can rarely be solved by just one person.

❞ We were inspired to ask for the public to help us by issuing public challenges. ❞ —John

John's team issued a challenge through the engineering website GrabCAD.com. Several teams submitted entries—some using strategies that John's team had not considered.

❞ You have to think differently when you're making something so enormous and so extremely lightweight. ❞ —John

John is not the only NIAC fellow using public contests to help form his design. Jonathan Sauder, an **engineer** at NASA's Jet **Propulsion** Laboratory, has hosted several contests to design components for his Hybrid Automaton Rover for Venus (HAR-V). Check it out in *Out of This World: Clockwork Venus Rover!*

Contest winner

The winning design, created by the Mexican engineer Abner Gómez, proposed a unique combination of inflatable parts and **tensegrity** structures to produce a lightweight **starshade.**

Tensegrity is short for *tensional integrity*. A structure that takes advantage of tensegrity is made up of two different elements: rigid *struts* (supports) and flexible cables. Think of it as something like an arrangement of rigid poles held together by a spiderweb of cables.

Gómez's design also includes inflatable elements. Think of a party noisemaker: Normally it is curled up, but it extends to form a rigid structure when someone blows into it. Balloons are also rigid when inflated.

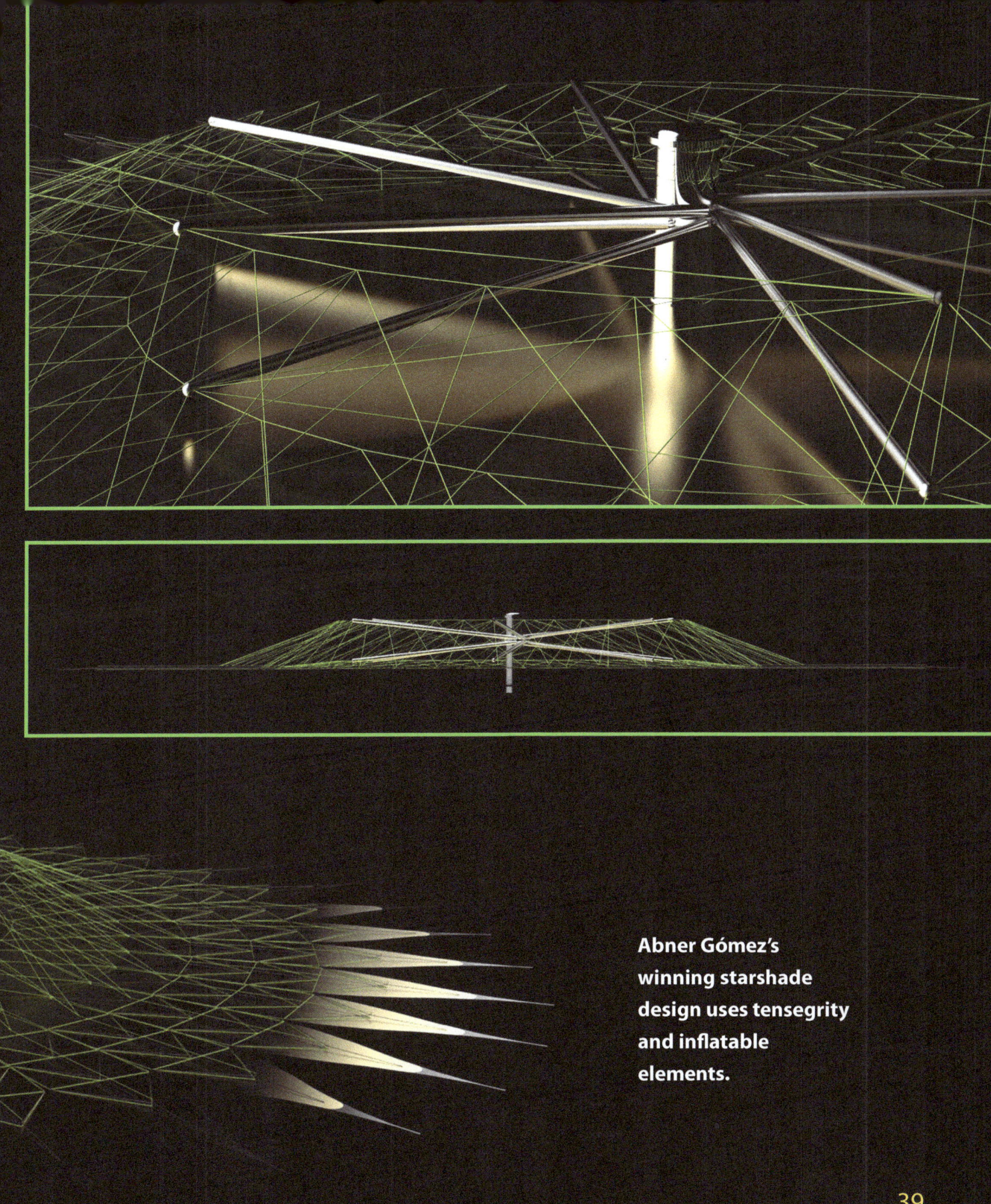

Abner Gómez's winning starshade design uses tensegrity and inflatable elements.

Other ideas

Other teams used different concepts that might be incorporated into the HOEE starshade design.

Some teams designed the petals with a slight bend to improve their stiffness. Think of a steel tape measure. It can be rolled up into a small coil. But it is fairly rigid when it is extended, due to the slight bend in its structure.

Other teams drew inspiration from the Eiffel Tower:

> ❝ People were shocked when the Eiffel Tower was built. How could you build something that was so tall and so light? ❞ —John

The secret to the Eiffel Tower's design is its use of *hierarchical structures*. A **truss** is a structure made up of rods or beams connected in a rectangular arrangement. The rods or beams are put together in such a way as to make the truss as a whole stronger. Now, what if you construct a larger truss made of smaller trusses, rather than rods or beams? That's what Gustav Eiffel did, and it is one way that teams approached the design challenge.

> ❝ You can make something extremely light, strong, and stiff if the only thing it has to hold up is itself. ❞ —John

Artist's rendering of the third-place winner, which uses trusses and inflatable technology

Future plans

> ❝ The next step is to go to the mechanical analysis that calculates the strength and stiffness and mass of all these ideas to see if they're good enough. ❞ —John

Despite the ingenuity of the designs, none of them were able to meet all the lofty design requirements. Next, John's team is partnering with the American Institute of Physics, a society for the advancement of physical sciences.

Undergraduate engineering students are invited to submit detailed **starshade** proposals and models of their designs. Students can win up to $10,000, and their ideas could be incorporated into the starshade that makes it to orbit.

Artist's concept of Kepler-186f, the first Earth-sized planet discovered in the habitable zone of its host star

Filling in the crossword puzzle of the universe

❝ We're working on the crossword puzzle of the universe. We're working on our little squares getting one letter at a time. This puzzle is immense, but we're making progress. There's a great satisfaction of contributing to it and once in a while getting the thrill of the discovery, that now we understand something that we didn't understand before. ❞ —John

Glossary

asteroid a rocky or metallic body smaller than a planet that orbits the sun.

atmosphere the mass of gases that surrounds a planet.

comet an icy body that releases gas or dust.

engineer a person who uses scientific principles to design structures, such as bridges and skyscrapers, machines, and all sorts of products.

exoplanet (extrasolar planet) a planet that orbits a star other than the sun.

gravitational pull also called gravitation or the force of gravity, the force of attraction that acts between all objects because of their mass. Because of gravitation, an object that is near Earth falls toward the surface of the planet. We experience this force on our bodies as our weight.

laser a device that produces a very powerful beam of light.

mass the amount of matter something contains.

orbit a looping path around an object in space; the condition of circling a massive object in space under the influence of the object's gravity.

payload bay the part of the rocket set aside for carrying cargo.

propellant solid or liquid fuel that is turned into gas and put under pressure to push a spacecraft forward.

propulsion pushing something, such as a spacecraft.

prototype a functional experimental model of an invention.

radiation energy given off in the form of waves or tiny particles of matter.

solar system the sun and everything that travels around it, including Earth and all the other planets and their moons.

starshade a shade used to block out the light from a distant star, allowing a telescope to see any planets around the star.

tensegrity a principal of architecture in which structures are supported by rigid struts, under compression, connected by flexible cables, under tension.

thrust moving force; a push with a force.

thruster a rocket or other device used to help steer and control the motion of a spacecraft.

truss a framework of beams or other supports usually connected in a series of triangles.

Review and reflect

Now that you've finished reading about John Mather, use these pages to think about his experiences and the Hybrid Observatory for Earth-like Exoplanets in new ways. As you work, reflect on the importance of creative problem solving, curiosity, and open-mindedness in life.

Complex problems and creative solutions

Why are astronomers interested in imaging Earth-like exoplanets?

What are some of the problems associated with imaging Earth-like exoplanets?

How does John Mather's Hybrid Observatory for Earth-like Exoplanets (HOEE) hope to overcome these challenges? What makes this solution so innovative?

Visit www.worldbook.com/resources to download sample answers, blank graphic organizers, and a rubric to evaluate writing.

Inspiration can come from anywhere!

Use a graphic organizer like the one below to map out your ideas. What ideas or experiences led to John's innovative solution?

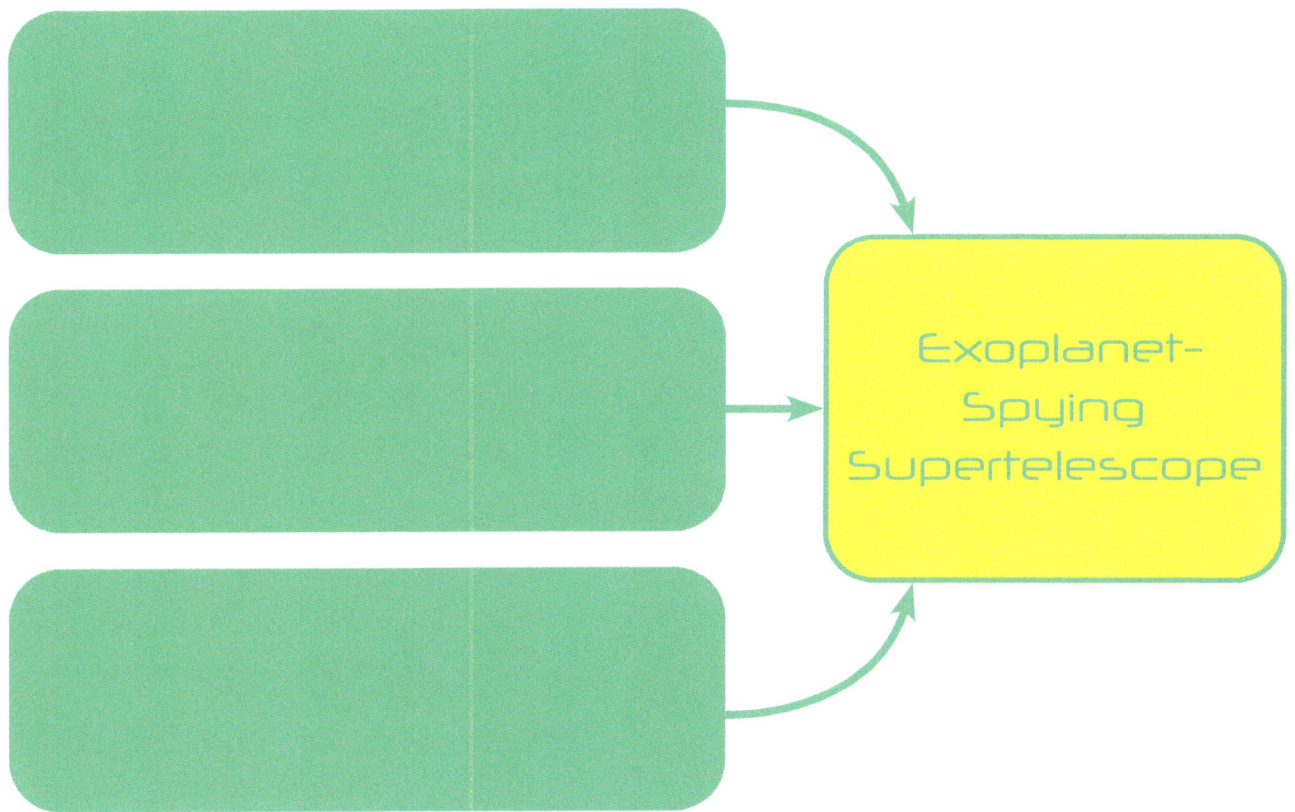

Write about it!

Think about John's experiences in life and as a NIAC Fellow.

How have John and his team used collaboration to develop HOEE? Why might it be important to work with others when looking for innovative solutions?

Index

A
adaptive optics, 30
American Institute of Physics, 42
astronomer, 11, 20, 25
astronomy, 26-27
atmosphere, 4, 17, 27, 30

B
big bang theory, 15

C
Cash, Webster, 21, 25
Chile, 26-27
contests, 36-37, 38, 42
Cosmic Background Explorer (COBE), 14-15, 18
cosmic microwave background (CMB) radiation, 12, 14-15

E
Eiffel, Gustav, 40
Eiffel Tower, 40
Einstein, Albert, 16
exoplanet, 6, 10-11
Extremely Large Telescope (ELT), 26, 31

G
Giant Magellan Telescope, 26-27
Gómez, Abner, 38
gravitational pull, 4
gravity, 11, 16

H
habitable zone, 4, 43
Hawking, Stephen, 18
Hybrid Observatory for Earth-like Exoplanets (HOEE), 29, 30, 31, 36, 38

I
ion propulsion, 35

J
James Webb Space Telescope (JWST), 22, 24-25, 28
Jet Propulsion Laboratory, 25, 37

K
Kepler-186f, 43

L
laser, 30

M
Mather, John, 6-7; boyhood, 8-9; COBE, 14-15; contests, 36; education, 12-13; JWST, 22; Nobel Prize, 18
Milky Way, 10
moon, 16-17

N
National Aeronautics and Space Administration (NASA), 13, 14, 25; NASA Innovative Advanced Concepts (NIAC), 7, 21, 37
Nobel Prize, 18

O
occultation, 16-17
occulting disk, 20
orbit, 32, 35

P
Penzias, Arno, 12

R
radiation. *See* cosmic microwave background (CMB) radiation

S
Sauder, Jonathan, 37
Smoot, George, 18
solar eclipse, 16-17
solar system, 25
Spitzer, Lyman, 20
starshade, 20-21, 28-29, 30, 32-33, 35, 36, 38, 40, 42
sun, 16-17

T
telescopes, 26-27: Extremely Large Telescope, 26, 31; Giant Magellan Telescope, 26-27; James Webb Space telescope, 22, 24-25, 28
tensegrity, 38
theory of general relativity, 16
thruster, 33
truss, 40

W
Wilson, Robert W., 12

www.ingramcontent.com/pod-product-compliance
Lightning Source LLC
Chambersburg PA
CBHW060936180426
43194CB00049B/2952